PRINCE2 2017® 2017 FOUNDATION REVISION NOTES	3
INTRODUCTION	3
Overview and basics	3
Project Characteristics	4
Integrated Elements of Prince2 2017	5
PRINCE2 2017 THEMES	6
Business case	6
Organization	7
Quality	7
Plans	7
Risk	7
Change	7
Progress	7
THE 7 THEMES – IN DETAIL	9
Business Case Theme	9
Organization Theme	10
Quality	14
Plans	16
Risk	18
Change	22
Progress	24
PROCESSES	26
Starting Up a Project (SU)	26
Directing a Project	27
Initiating a Project	28
Controlling a stage	30
Managing Product Delivery	30
Managing a Stage Boundary	31
Closing a Project	32
PRINCE2 2017 – FOUNDATION EXAM - GUIDELINES	34
PRINCE2 2017 – FOUNDATION EXAM STRATEGY	36
PRINCE2 2017 – PRACTITIONER EXAM - GUIDELINES	38
TABBING THE MANUAL	41
PRINCE2 2017 - THINGS TO ADD TO THE MANUAL	42
Themes	42
Business Case Theme	42
Organisation Theme	42
Quality Theme	43
Plans Theme	43
Risk Theme	44
Change Theme	44
Progress Theme	44
Processes	45
Starting Up a Project Process (SU)	45
Directing a Project Process (DP)	46
Initiating a Project Process (IP)	46

CONTROLLING A STAGE PROCESS (CS) 46
MANAGING A PRODUCT DELIVERY (MP) 47
MANAGING A STAGE BOUNDARY PROCESS (SB) 47
CLOSING A PROJECT PROCESS (CP) 47

PRINCE2 2017® 2017 Foundation Revision Notes
Introduction
This document is intended as a revision guide\study aid and is not intended to be a replacement for either dedicated revision or a structured course

It is designed to supplement this training and revision. It provides the key points in each area of the exam and the focus areas. I created this to supplement the training materials I use in my daily role which is that of an independent trainer of Prince2 2017 and other Project Management methodologies

The revision notes are broken into the following sections:

1. Foundation Exam
    a. Overview and basics
    b. Prince2 2017 Principles
    c. Prince2 2017 Themes
        i. Minimum Requirements
        ii. Themes in detail
    d. Prince2 2017 Process
    e. Foundation Exam Guidelines
    f. Foundation Exam Strategy
2. Practitioner Exam
    a. Practitioner Exam Guidelines

Overview and basics
The 7 Principles
- Continued Business Justification
- Learn from Experience
- Defined Roles and Responsibilities
- Manage by Stages
- Manage by Exception
- Focus on Products
- Tailor to suit the Project environment

The 7 Themes
- Business Case
- Organisation
- Quality
- Plans

- Risk
- Change
- Progress

The 7 Processes
- Starting Up a Project Process (**SU**)
- Directing a Project Process (**DP**)
- Initiating a Project Process (**IP**)
- Controlling a Stage Process (**CS**)
- Managing Product Delivery Process (**MP**)
- Managing a Stage Boundary Process (**SB**)
- Closing a Project Process (**CP**)

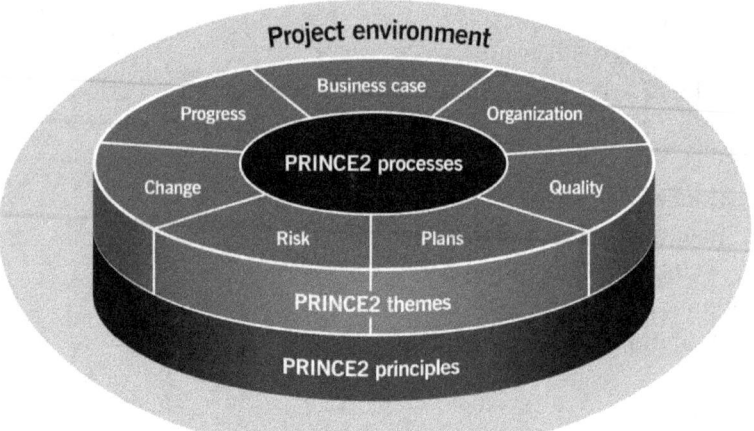

Figure 1 - Prince2 2017 Integrated Elements

Project Characteristics

The characteristics of a Prince2 2017 Project are:
- **Change** – A means by which we implement change or introduce change into an organisation
- **Temporary** – Prince2 2017 projects have a defined start and End
- **Cross-Functional** – Different skills are used to deliver a project, some skills may have the same title in BAU, some skills may come from internal or external suppliers or it may be resourced within a mix of both
- **Unique** – Each project is unique – no project is the same ever, it can be a different Exec, different product or different organisation
- **Uncertain** – Projects are riskier than Business as Usual (BAU) – this is an inherent character of a project

Within Prince2 2017, there are 6 objectives relating to project performance that need to be managed and monitored by the Project Manager as throughout the projects lifecycle, these are:
- Time
- Cost
- Quality
- Scope
- Benefits
- Risk

These are also known as:
- Variable
- Performance targets
- Aspects of Project Management

Integrated Elements of Prince2 2017

The Prince2 2017 methodology is made up of 4 elements or components, these are known as the **4 integrated elements** that when combined enable and empower the Project Manager, these are:
- **The 7 Principles** – Provide Guidance and good/best practice – the Principles enable Prince2 2017 to be a generic methodology
- **The 7 Themes** – Are aspects of project management that are addressed continually throughout the project and are linked to the 7 Principles (for example – The Business Case is directly linked to the Continued Business Justification Principle)
- **The 7 Processes** – The sequential steps that are followed by the Project Manager throughout the project lifecycle
- **Tailoring** (Project environment) - Appropriate use of the framework based upon the organisation and the projects complexity

It is important to remember them in this order as this is the order of importance in Prince2 2017, the Principles are the underpinning best practice that enables the method to be truly generic, the Themes are aspects or functions that the Project Manager will observe, practice, undertake through the projects entire lifecycle, the Processes are the sequential steps to deliver the project from start to closure and the Tailoring is undertaken to ensure the project is delivered in a manner appropriate to Prince2 2017, the project based upon the complexity and risk associated and finally the organizational standards

*** Exam Tips ***
Principles, Themes and Processes
Within the Prince2 2017 217 exam, you will occasionally get the following question in the exam:
- Which of the following is a Principle?" or "Which of the following is NOT a Principle?
- Which of the following is a Theme?" or "Which of the following is NOT a Theme?
- Which of the following is a Process?" or "Which of the following is NOT a Process?

It is **NOT** necessary to memorise all pf 7 Principles, 7 Themes and the 7 Processes, simply apply these 3 simple rules to be able to pick the right answer from a list:

- **Principles are statements** – For example, it is: *LEARN* from experience, is it NOT Learning from experience
- **Processes start with a verb that ends "ing"** – For example: Starting up a project
- **With the exception of the Business Case Themes are all one word**

Prince2 2017 Themes
A new part of the PRINCE2 2017 version is the introduction of Minimum requirements for use with each of the 7 Themes, it is now very common that they test you on these in the foundation exam, and from experience there have been an average of 6 questions in recent Foundation exams I have been the examiner for. So these are important to understand

The guide below is a useful revision aid to gain a better understanding of the Minimum Requirements relating to the Prince2 2017 Themes

Individual Theme Minimum requirements
Business case
- Create and maintain a business justification for the project (In some form of document)
- Review and update the business justification throughout the project's lifecycle
- Define management actions that will be put in place to ensure that the project's outcomes are achieved and confirm that the project's benefits are realized

PM Guide

- Define and Document the roles and responsibilities for the Business Case and Benefits Management

Organization
- Define its organization structure and roles.
- Define the rules for any delegated change authority
- Define its approach for communicating and engaging stakeholders

Quality
- Define its quality management approach
- Specify explicit quality criteria for products in their product descriptions
- Maintain records of planned quality activities that have been carried out
- Specify the customer's quality expectations and prioritized acceptance criteria for the project
- Use lessons to inform quality planning

Plans
- Ensure that plans enable the business case to be realized
- Have at least two management stages: an initiation stage and at least one further management stage
- Produce a project plan for the project as a whole and a stage plan for each management stage
- Use a product-based planning approach for the project plan, stage plans and exception plans
- Produce specific plans for managing exceptions
- Define the roles and responsibilities for planning
- Use lessons to inform planning

Risk
- Define its risk management approach
- Maintain some form of risk register
- Ensure that project risks management takes place throughout the project lifecycle
- Use lessons to inform risk management

Change
- Define its change control approach
- Define how product baselines are created, maintained and controlled
- Maintain some form of issue register
- Ensure that project issues are management throughout the project lifecycle
- Use lessons to inform issue management

Progress
- Define its approach to controlling progress in the PID
- Be managed by stages

PM Guide

- Set tolerances and be managed by exception
- Review the business justification when exceptions are raised
- Learn lessons

The 7 Themes – In Detail
Business Case Theme

Purpose

The Business Case is used to judge whether a project is (and remains) desirable, viable and achievable.

Basics, Key word: "Justification"
Owned (Responsibility of) the EXECTUTIVE
Directly aligned to the Continued Business Justification Principle

Minimum requirements
- Create and maintain a business justification for the project
- Review and update the business justification throughout the project
- Define management actions that will be put in place to ensure that the project's
- outcomes are achieved and confirm that the project's benefits are realised
- Define and Document the roles and responsibilities for the Business Case and
- Benefits Management

Development path
- Outline Business Case created in SU – Developed from the Project Mandate
- Detailed Business Case created in IP – Developed from the Outline Business Case and the Project Brief
- Updated during the SB process – in line with the Continued Business Justification Principle
- Review throughout the projects lifecycle (CS Process) when assessing and managing
    - Risks – that could have an effect upon the projects objectives
    - Issues – that could affect the projects objectives

Other key points

Each Business Case should have a minimum of 3 options within a Prince2 2017 project
Do nothing' should always be the starting option to act as the basis for quantifying the other options. The difference between 'do nothing' and 'do the minimum' or 'do something' is the benefit that the investment will buy.

PM Guide

Key Products
Business Case - Appendix A.2 – P292

A business case is used to document the business justification for undertaking a project, based on the estimated costs (of development, implementation and incremental ongoing operations and maintenance costs) against the anticipated benefits to be gained and offset by any associated risks. It should outline how and when the anticipated benefits can be measured.

- Documents the Justification for the project (supported by the Benefits Management Approach)

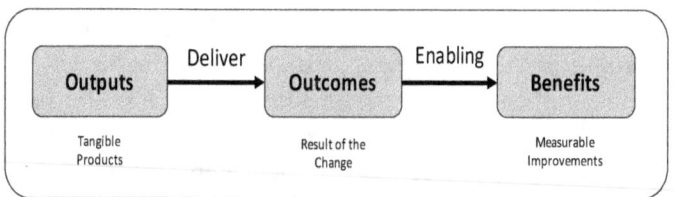

Figure 2 - Outputs, Outcomes & Benefits

Prince2 2017 projects deliver products, these are the purpose or reason for the projects existence and called Specialist Products. These products are also known as Outputs

The production or creation of the Outputs, delivers the Outcomes or the business change, which in turn enables the realisation of the benefits (as shown above)

Benefits Management Approach – Appendix A.1 – P294
- Defines any management actions needed to ensure benefits are realised
- Describe how and when benefits that have actually accrued will be measures
- Created at the same time as the Detailed Business Case (IP)
- Updated at the same time (SB)
- Passed-on to Corp/Prog. management during CP process
- The only document to remain active post project

Organization Theme
Purpose
Defines and established the projects structure of accountability and responsibility. (popular answer or question) it is essentially answering the questions:

- Who is involved in the project
- What is their role
- What level of authority do they have

Basics
Each project must have the primary roles or interests represented, Business, User and Supplier within the Project Board

One person can have more than one role, it is key to understand that these are roles and not people

You may need more than one person to cover some roles characteristics

Minimum requirements
- Define its organization structure and roles.
- Define the rules for any delegated change authority
- Define its approach for communicating and engaging stakeholders
- Details of specific roles

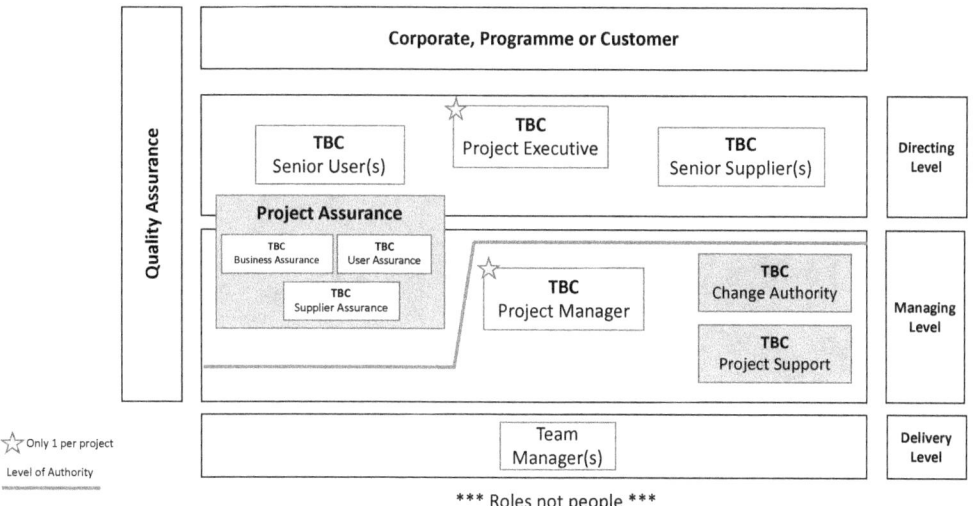

Figure 3 - Project Management Team Structure

Project Management Team Structure
The Project Board within a Prince2 2017 project has a set of defined roles, again which allow the flexibility to support any type of project, these roles or the

PM Guide

responsibilities for these must be undertaken within the Project Board. As mentioned above, these roles can be combined or shared according to the project's needs, however it is crucial that these roles are always allocated

The 3 roles or responsibilities are the Business, User and Supplier which combined make up the Project Board

Project Board
Decision making – The Directing a Project process covers the work of the Project Board

Executive
- Represents the Business Interest
- Ultimate Decision maker (Ultimately accountable for project success or failure)
- MUST be an INDIVIDUAL
- Provides funding for the project
- Owns Business Case

Senior User(s)
- Represents End users and Impacted Users
- Accountable for Benefits realisation
- Defines Customer Quality Expectations and Acceptance Criteria for the project

Senior Supplier(s)
- Represents supplier (Specialist) interests
- Accountable for the delivery of Quality
- Commits supplier (Specialist) teams or resources

Change Authority
- Approves/Rejects Requests for Change or Off-Specifications
- The Change Authority reduces the number of Requests for Change that need to go to the Project Board

PM Guide

Project Manager
- Responsible to the Day to Day management of the project
- Manages Risks and Issues that are escalated to them
- Sets-up project procedures

Project Support
- Support the Project Manager (May also support the Team Manager)
- Popular activities may include: Config Mgnt, Doc Control, Admin, maintaining of
- registers)
- Responsible for the Quality Register
- Reviews Checkpoint reports to support the information contained within the Highlight Report
- Creates the Product Status Account for End Stage Reports\Assessments, End Project Reports\Assessments or to support Exception Plans
- Sometimes performed by Corporate/Programme management (Project management office)

POPULAR FOUNDATION QUESTIONS
- Question: Which role is sometimes performed by Corporate/programme management?
- Answer: Project support

Team Manager
- Works is covered by the Managing Product Delivery process
- Manages the work of a work package
- Provides progress updates (Checkpoint Reports) to the Project Manager

Minimum requirements
- Define its organization structure and roles.
- Define the rules for any delegated change authority
- Define its approach for communicating and engaging stakeholders

Key Products
Communication Management Approach – Appendix A.5 – P299
A communication management approach contains a description of the means (procedures, tools, techniques) **and frequency of communication with parties both internal and external to the project.**

PM Guide

Quality
Purpose
Defines the means by which the project will create and verify products that are Fit for purpose and meet their requirements.

Basics
The right level of quality is defined as something that is Fit for Purpose.

Minimum requirements
- Define its quality management approach
- Specify explicit quality criteria for products in their product descriptions
- Maintain records of planned quality activities that have been carried out
- Specify the customer's quality expectations and prioritized acceptance criteria for
- the project
- Use lessons to inform quality planning

4 Key elements of Quality
Quality Planning
- Understanding the level of quality needed and how it will be managed

Quality Control
- Reviewing and testing
- Quality Review Technique used

Quality Management Systems
- A set of Corporate policies and standards – External to the Project

Quality Assurance
- External to the project (The Responsibility of Corporate/Programme Management)

Quality Review Techniques
- A technique to Assess conformity of a product against its Product Description
- A technique to sign-off or Baseline a product
- This is NOT used for users to add or change a product (That would be a Request for Change)

Quality Review Types

Within a Prince2 2017 project there are 2 types of reviews within the Quality Theme, In-Process and Appraisal

Each of these must be defined within the Project Approach, to understand the project's delivery method

- In-Process – In-Process appraisal methods are planned during the creation of the projects specialist products and are built into the planned delivery
- Appraisal - Appraisal methods are used to assess the final products for its completeness and assess whether it is fit for purpose and meets the customers quality expectations

Roles involved
- Chair – Responsible for the conduct of the review
- Reviewer(s) - Person or people that will review/test the product
- Administrator – Supports the chair (Minutes/actions)
- Presenter – Represents the person/team that created the product being review

Key Products
Project Product Description – Appendix A.21 – P322
- A description of the final solution and the acceptance criteria
- Created in Starting up a Project process
- Refined in Initiating a Project Process
- Reviewed at each Stage Boundary

Key headings: "Customer Quality Expectation" and "Acceptance Criteria" – provided and owned by Senior User or representatives of the Senior User
- Customer Quality Expectation – In early stages. often expressed in broad terms to get a common understanding of general quality requirements
- Acceptance Criteria – A specific list of measurable criteria that a project product should
- meet in order to be acceptable

Product Description – Appendix A.17   P315
A description of a particular product that the project or stage needs to create that will support the realisation the projects benefits

PM Guide

Quality Management Approach – Appendix A.22 – P324
Describes Procedures, tools and techniques with regards to Quality

Quality Register
Keeps a track of the dates and results of any planned quality checks

Plans
Purpose
Defines the means for delivering products (where, how, whom, when and how much).

Minimum requirements
- Ensure that plans enable the business case to be realized
- Have at least two management stages: an initiation stage and at least one further management stage
- Produce a project plan for the project as a whole and a stage plan for each management stage
- Use a product-based planning approach for the project plan, stage plans and exception plans
- Produce specific plans for managing exceptions
- Define the roles and responsibilities for planning
- Use lessons to inform planning

Levels of plans
PRINCE2 2017 recommends different levels of plans to cater for the different levels of detail
required by the different levels of Management

Within a Prince2 2017 project, there are 3 levels of plan

Recommended Levels of Plans
- Project Plan – Appendix A.16 – P313 (used for all levels of plan)
- Stage Plan
- Team Plan

The same product description is used for all plans and simply refined

16

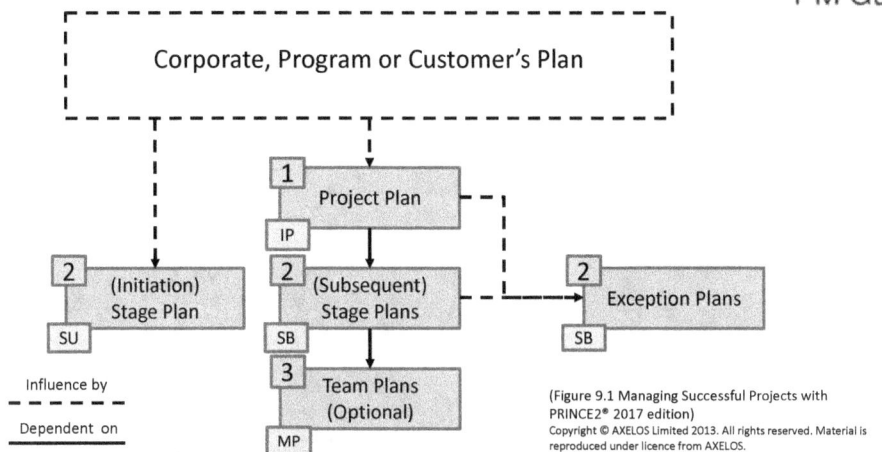

Figure 4 - Levels of Plan within Prince2 2017

Important - The Team plan is an *OPTIONAL* plan, it is therefore recommended, however, if you get asked the questions about which plans are MANDATORY, that would only be Project and Stage

Important: An Exception plan is NOT a level of plan, it is a replacement plan for either the Project or a Stage, whichever one of them is in Exception and requires re-planning, when approved the Exception Plan replaces the affected plan

Plan Procedure
- Designing the plan
- Defining and analysing the products
- Identifying activities and dependencies
- Preparing estimates
- Preparing the schedule
- Documenting the Plan
- Analysing risks to a plan

Product Based-Planning
1. Write the Project Product Description (First step – This is a popular question)
2. Create Product Break-down Structure
3. Write Product Descriptions
4. Create Product Flow Diagram

It is important to remember that the Product Based Planning technique is a sub process of the Defining and Analysing Products – this is a representation of the process in full below unlike how it is represented in the official manual

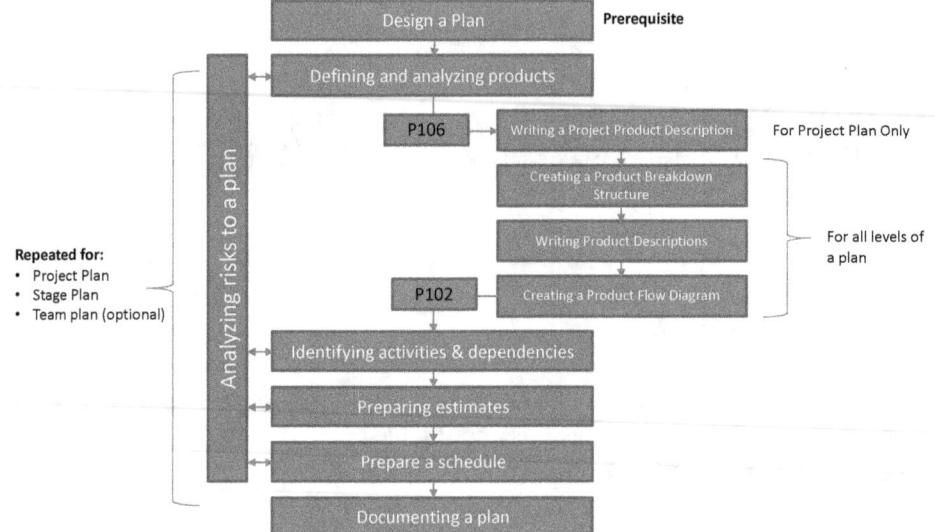

Figure 5 – Prince2 2017 – Guide to effective planning

Key Products
Product Break-down Structure (PBS) – Appendix D – P350-356
- A hierarchical breakdown of the projects main products or outputs

Product Flow Diagram (PFD) -Appendix D – P356
- A sequence in which products need to be created

Risk
Purpose
Identify, assess and control uncertainty. The Prince2 2017 definition of risk management is:

"The systematic application of principles, approaches and processes to the tasks of identifying and assessing risks, planning and implementing risk responses and communicating risk management activities with stakeholders"

Basics
- Key word: Uncertainty "If, May, Could, Might" – all of these keywords indicate an event that "may happen". A risk is something that has not yet occurred
- Supports the principle Continued Business Justification
- The assigned roles & responsibilities for risk management (Risk Owner and Risk Assignee) supports the principle of Defined Roles & Responsibilities

Minimum requirements
- Define its risk management approach
- Maintain some form of risk register
- Ensure that project risks management takes place throughout the project lifecycle
- Use lessons to inform risk management

Risk Appetite
An organizations unique approach to risk taking will determine the projects attitude to risk taking and help in defining the risk tolerances

Risk Budget
Risk budget is an optional budget agreed during the Initiating a Project Process and is "ring-fenced" for the implementation of risk responses
- Actions to reduce the probability or impact of an identified risk

And not to be used for any other purpose, in the event it is not used, it is to be returned to Corp, Prog or Customer for reallocation

Risk Procedure
- Identify
- Assess
- Plan
- Implement
- Communicate

Identify
- Context: The project objective that is at risk
- SU: This will go in the Daily Log
- IP onwards: This will go in the Risk Register

Description:
- *CAUSE*: The Source of the risk or A known event that has brought light to a risk
- *EVENT*: The area of uncertainty in terms of the threat or opportunity
- *EFFECT*: The impact on the project objectives should the risk materialise

Assess
- Estimate: Assess and understand the individual risk
- Probability: Likelihood of the risk occurring
- Impact: how it would affect objective
- The impact of the risk upon the Project Plan, Stage Plan and Business Case
- Proximity: How close (in time) is the risk or how quickly it is likely to occur
- Evaluate: Aggregate (Total) exposure to risk

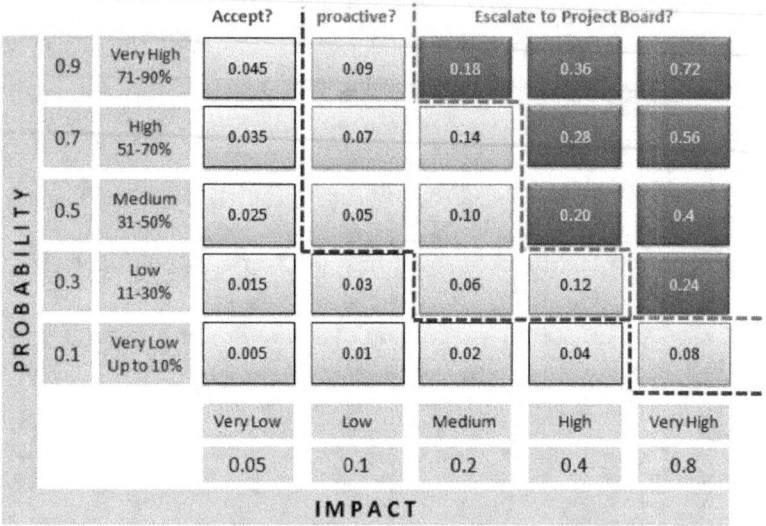

Figure 6 - Prince2 2017 Summary Risk Profile

Plan
The plan step involves assessing and choosing the appropriate response to the identified risk, primarily aimed at removing either the probability or impact of the risk

PM Guide

Threat	Opportunity
Avoid	Exploit
Reduce	Enhance
Transfer	
Share	
Accept	
Contingency Plans	

Each of the risk responses should be assessed and the most appropriate response selected

Implement
The chosen risk response needs to be actioned and the effectiveness of the response monitored, where needed corrective actions taken to further control the risk

The risk responses need to be managed using identified Owner and Actionee for each risk:

- Risk Owner: Someone that will monitor and report back on an individual risk assigned to them
- Risk Actionee: Someone that will implement a risk response

Communicate
- Risk communication should happen throughout the management of a risk (As per stakeholder requirements in the Communication Management Strategy)
- Risk communication is part of the following management products:
    - Checkpoint reports
    - Highlight reports
    - End stage reports
    - End project reports
    - Exception reports

PM Guide

Key Products

Risk Management Approach – Appendix A.24 - P327
Describes Procedures, tools and techniques with regards to Risk Management
Includes an organization Attitude to risk taking (Appetite – Tolerance)

It does NOT include the Risk Budget (This is often put as an option in the answers but is WRONG) The Risk Budget is part of the Project Plan and the Project Controls

Risk Register – Appendix A.25 - P329
Used to keep a record and track on risks that are being managed

Change
Purpose
Identify, assess and control and potential and approved chance to baselines.

Minimum requirements
- Define its change control approach
- Define how product baselines are created, maintained and controlled
- Maintain some form of issue register
- Ensure that project issues are management throughout the project lifecycle
- use lessons to inform issue management

Types of Issues
Within Prince2 2017, all changes are classified as issues and there are only 3 types of issue:

Request for change
- Someone asking for something different or a proposed change to the baseline

Off-Specification
- Where something has been delivered that doesn't meet the Product description (criteria or specification) asked of it

Problem or Concern
- Any other issue

Handling the Issues
Issue and change control procedure
1. Capture
2. Assess
3. Propose
4. Decide
5. Implement

Request for Change or Off-Specification
These go to the Change Authority for assessment and review of options and recommendation actions

Off-Specifications are essentially products that have been delivered or will be delivered by the supplier that do not meet the agreed specifications as agreed by the Work Package. The important point is to understand that the remediation of an Off-Specification

An agreement to an Off-Specification is a concession, and any concession leads to a Request for Change to the baselined products to align the product description or associated documents to the new specification

Problem or Concern:
- Informal issues go in the Daily Log
- Formal issues go in the Issue Register for management

Key Products
Change Control Approach – Appendix A. - P
Describes Procedures, tools and techniques with regards to Issue/change management and forms part of the Project Initiation Documentation (PID)

Issues Register – Appendix A. - P
Used to record and keep track of any issue that needs to me managed

Issues Report – Appendix A. - P
Provides information on an issue (Description, Impact, Options)

Daily Log – Appendix A. - P
Used to capture Informal issue that do NOT need to be managed formally

PM Guide

Configuration item record – Appendix A. - P
Includes Name, version, status of products, and important relationships between items

Product status report – Appendix A. - P
An Ad-hoc report created to understand the status of a group of products within the project, usually within defined parameters – essentially a report on the status of all products being created by the project

Progress
Purpose
Establish mechanisms to monitor actual achievements against those planned; and control any unacceptable deviations.

Minimum requirements
- Define its approach to controlling progress in the PID
- Be managed by stages
- Set tolerances and be managed by exception
- Review the business justification when exceptions are raised
- Learn lessons

Supports the following principles:
- Manage by exception (Popular question/answer)
- Manage by stages

Tolerance
- A permissible deviation above or below a plans target for time and cost before the need to escalate

Exception
- Where it is FORECAST that a tolerance will be exceeded

Controls
There are only 2 Time-driven controls in PRINCE2 2017:
- Highlight Report
- Checkpoint Report

All other controls are event driven – an event is the trigger for the report, for example the End Stage Report is triggered by the end of the active stage

Stage Boundary
The purpose of the managing a stage boundary process is to enable the project manager to provide the project board with sufficient information to be able to:
- Review the success of the current management stage
- Approve the next stage plan
- Review the updated project plan
- Confirm continued business justification and acceptability of the risks.

Effectively provides a Go/No go decision point for the Project Board

Management\Delivery Stages (these are the terms used within Prince2 2017)
A commitment of resources with a go/no go decision point in between

The Startup Stage and Initiation Stage are often called Management Stage whilst the subsequent stages are called Delivery Stages

Technical Stage (NOT part of Prince2 2017)
A piece of work that requires specific or specialist skills (although Prince2 2017 recognizes technical stages and agrees that they need to be considered and included in a project, Prince2 2017 does NOT include how to manage them in the framework as they are specific to a project, sector or industry)

Key Products
Highlight Report - – Appendix A. - P
Progress report provide from the Project Manager (CS) to the Project Board (DP)
These are provided at a frequency agreed as part of the Communication Management Strategy (This is a popular question/answer)

Checkpoint Report – Appendix A.4 - P298
Progress report provide from the Team Manager (MP) to the Project Manager (CS)
These are provided at a frequency agreed when agreeing a work package

End Stage Report– Appendix A.9 – P303
Created as part of the Managing a Stage Boundary process
Provides confirmation on what was completed in the stage that we are coming to the end of and a progress of the project so far

End Project Report– Appendix A.8 – P301
Provides confirmation of what was delivered by the project, along with an evaluation of the project

Processes
Starting Up a Project (SU)
Purpose
Provide the Project Board with sufficient information to make an informed decision that the project is both viable and worthwhile
Ensure that the pre-requisites for Initiating a project are in place and any blockers to a successful initiation are removed or highlighted as risks
It is also to prevent poorly conceived projects from even being initiated

Base Information
- Classed as Pre-project or feasibility
- It is triggered by the Project Mandate supplied by Corp, Prog or Customer
- It's the first process in Prince2 2017
- Directing a Project (DP) is the Process that follows SU (Popular question/answer) and not Managing a Stage Boundary (SB)

Key Products
Project Brief – Appendix A.19 – P317
A project brief is used to provide a full and firm foundation for the initiation of the project and is created in the Starting Up a Project Process and Describes the purpose, cost, time, performance requirements and constraints for the project

The contents of the Project Brief are:
   Project Definition -
   Explains what the project needs to achieve. It should include:
   - Background
   - Project objectives (covering time, cost, quality, scope, benefits and risk performance)
   - Desired outcomes
   - Project scope and exclusions
   - Constraints and assumptions
   - Project tolerances
   - The user(s) and any other known interested parties
   - Interfaces

Outline Business Case – Appendix A.2 – P.
- Reasons why the project is needed and the business option selected. This will later be developed into a detailed business case during the initiating a project process

Project Product Description - Appendix A.21 - P
- Includes the customer's quality expectations, user acceptance criteria, and operations and maintenance acceptance criteria

Project Approach
- Defines the choice of solution that will be used within the project to deliver the business option selected from the business case. This will take into consideration the operational environment into which the solution must fit and any tailoring requirements (if known)

Project Management Team Structure
- A chart showing who will be involved with the project

Role descriptions – Appendix C
- These describe the roles of those in the project management team and any other key resources identified at this time

References
- These include references to any associated documents or products.

Directing a Project
Basics
Process starts after Starting Up a Project (Popular question/answer)

Purpose
- To enable the Project Board to be accountable, by giving them control with key decisions, whilst also freeing up their time by delegating the day-to-day management to the Project Manager

Other key information
Covers the work of the Project Board (Popular quest/answer)
"Provide Ad-Hoc direction" is the activity that receives Highlight reports or exception reports (This is sometimes a question/answer)

PM Guide

Keywords: Authorise, Approve, Direction, Advice

Initiating a Project
Purpose
- Set solid foundations for the project
- Understand more about what is needed to deliver the project before any significant spend in committed (or any resources committed)
- Defines how the project will be managed, controlled and tailored
- Defines the structure of the Project Management Team

Objectives
To summarize how Prince2 2017 will be tailored for the project

Key Products
Project Initiation Documentation (P.I.D.) – Appendix A.20 - P319
Forms the 'Contract' between the Project Manager and Project Board confirming how the project will be managed and control and is reviewed as part of the Managing Stage Boundary Process (SB)

Project Definition
Explains what the project needs to achieve. It should include:
- Background
- project objectives and desired outcomes
- project scope and exclusions
- constraints and assumptions
- the user(s) and any other known interested parties
- interfaces

Project Approach
- Defines the choice of solution that will be used in the project to deliver the business option selected from the business case, taking into consideration the operational environment into which the solution must fit

Business Case – Appendix A.2 – P294
- Describes the justification for the project based on estimated costs, risks and benefits

28

Project Management Team Structure
- A chart showing who will be involved with the project

Role descriptions
- These describe the roles of those in the project management team and any other key resources

Quality Management Approach – Appendix A.22 – P324
- Describes the quality techniques and standards to be applied, and the responsibilities for achieving the required quality levels. Where the project is subject to the commissioning organization's quality management policies/strategies, the PID should make reference to
- them rather than duplicate them. Where the project is not subject to the commissioning organization's quality management policies/strategies, appropriate strategies/approaches should be documented

Change Control Approach - Appendix - A.3 – P 296
- Describes how and by whom the project's products will be controlled and protected. Where the project is subject to the commissioning organization's change control policies/strategies, the PID should make reference to them rather than duplicate them. Where the project is not subject to the commissioning organization's change control policies/strategies, appropriate strategies/approaches should be documented

Risk management approach – Appendix A.24 – P327
- Describes the specific risk management techniques and standards to be applied, and the responsibilities for achieving an effective risk management procedure. Where the project is subject to the commissioning organization's risk management policies/strategies, the PID should refer to rather than duplicate them. Where the project is not subject to the commissioning organization's risk management policies/strategies, appropriate strategies/approaches should be documented

Communication management approach – Appendix - A.5 – P299
- Defines the parties interested in the project and the means and frequency of communication between them and the project. Where the project is subject to the

commissioning organization's communication management policies/strategies, the PID should make reference to them rather than duplicate them. Where the project is not subject to the commissioning organization's communication management policies/strategies, appropriate strategies/approaches should be documented

Plan – Appendix - A.16 – P311
- Describes how and when the project's objectives are to be achieved, by showing the major products, activities and resources required on the project. It provides a baseline against which to monitor the project's progress, management stage by management stage

Project Controls
- Summarizes the project-level controls such as management stage boundaries, agreed tolerances, monitoring and reporting

Tailoring of PRINCE2
- A summary of how PRINCE2 will be tailored for the project.

Controlling a stage
Purpose
Covers the Day-to-Day work of the Project Manager through the delivery of the project
Focus here is on the Project Manager delivering the stages products to the right level of quality on-time and within budget by:
- Assigning work (issuing Work packages)
- Monitoring work (Receiving Checkpoint reports and updating the Stage Plan)
- Managing risks/Issues (and escalating if needed)
- Reporting progress (Providing Highlight Reports to the Project Board)

Objective
Focus on delivering the stages products

Managing Product Delivery
Basics
Each specialist product delivered is created\purchased\built within the Managing Product Delivery Process

Purpose
- Provides a controlled link between the Team Manager and the Project Manager
- Team Manager: accepts the works, gets it done and hands it back

The process of accepting a Work Package is essentially a negotiation between the Project Manager and Team Manager and once agreed, the Work Package is accepted and the Project Manager authorizes the delivery

Key Products
Work Package – Appendix A.26 – P330
A work package is a set of information about one or more required products collated by the project manager to pass responsibility for work or delivery formally to a team manager or team member.

Checkpoint Report – Appendix A. 4 – P298
A checkpoint report is used to report, at a frequency defined in the work package, the status of the work package

Managing a Stage Boundary
Basics
Takes place towards the end of a Stage or at the Request for an Exception plan

Purpose
- Review the stage that we are coming to the end of and prepare for the next stage
- To enable the Project Board to be provided with the necessary information to approve the next stage
  - **NOTE**: The approval is NOT part of this Process. The approval is part of the DP process

Other key Points
- The Managing a Stage Boundary process is also used for the creation of an Exception Plan
- The Managing a Stage Boundary process can be triggered from the following processes:
    - Initiating a project (Towards the end of the Initiation stage)
    - Controlling a Stage (Towards the end of a management stage)

PM Guide

- Directing a Project (From the request for an Exception plan)

Key products
End Stage Report – Appendix A.9 – P303
Information to be presented to the Project Board about the project performance during the stage and the project status at stage end

Stage Plan – Appendix A.16 – P311
Detailed plan for the next stage

Closing a Project
Basics
Takes place towards the end of the final stage or if a premature closure to the project is requested

Purpose
Provide a fixed point for which the project solution is handed over into Business as Usual (BAU) and the Project team is disbanded.

Other Key Points
- Pass-on any open risks
- Pass-on the Benefits Review Plan
- Evaluate the Project (Against v1.0 of the P.I.D and how/why it changed throughout the project)
- Archive project information (To enable future audits)

Key products
End Project Report
- An end project report is used during project closure to review how the project performed against the version of the PID used to authorize it. It also allows the passing on of:
    - Any lessons that can be usefully applied to other projects
    - Details of unfinished work, ongoing risks or potential product modifications to the group charged with future
- support of the project's products in their operational life.

Lessons Report - Appendix A.15 – P312
Defines learning points of how future project can be improved – The lessons report

Draft Closure Notification (Should include final dates for costs to be charged to the project)

Prince2 2017 – Foundation Exam - Guidelines

Introduction
This guide is provided to give delegates an understanding of the Prince2 2017 Foundation exam paper:
- Structure
- Topics
- Types of questions

Exam Structure
- Simple Multiple-choice
- 1 hour duration
- 60 Questions
- Pass Mark 55%
- Closed Book

Looking at these points in a little more detail

Simple Multiple-Choice
The PRINCE2 2017 foundation exam is a simple multiple-choice exam, this means that each question will have 4 options for answers and only 1 correct answer.

1 hour in duration
The exam is a 1 hour exam, We will discuss the exam strategy and time management further on.

60 Questions
The exam will have 60 multiple-choice questions.

Pass mark 55%
The pass mark for the exam is 55%, therefore, you will need to score 33 out of 60 that are mark (The 5 Trial/Feedback questions will NOT be marked) to be awarded a pass in the exam

Closed Book
The exam is a closed book exam, this means that you are NOT allowed any additional support material in the exam with you.

Topics covered

The objective of the examination is to enable a candidate to demonstrate an understanding of the PRINCE2 2017:
- Principles
- Processes
- Themes
- Techniques
- And Roles

The exam paper will NOT be broken up into sections, the questions on the topic areas above will be scattered around the paper in what would seem, no particular order.

Types of questions

Within the PRINCE2 2017 foundation exam you may encounter the following different types of multiple-choice questions:
- Standard
- Negative
- Missing word
- List
- Standard multiple-choice

The first style of question that you might see, is a standard multiple choice question, here you will see that there are 4 possible answers provided and only one correct answer.

Negative multiple-choice

In this style of question, your thought process is reversed, as you are now looking for the negative, so in this example, we are looking for the answer that is NOT a key element that needs to be balanced when defining management stages

Try not to get caught out by these, If you get this style of question, the **NOT**, will always be in bold text.

Missing word

In this style of question, you need to replace the square brackets and question mark with the word that best fits the sentence.

PM Guide

Prince2 2017 – Foundation Exam Strategy

Introduction

This guide aims to provide delegates with crucial advice on how to tackle the Prince2 2017 Foundation exam, in this guide we will cover the following areas:

- Exam Strategy
- Time Management
- Question Techniques

Exam Strategy
Strategy 1 – Start to Finish
- Quite simply, the first strategy is to start at question 1 and answer each question in turn
- until you come to the end of the paper.
- If time is not a problem for you, this could be the strategy for you.

Strategy 2 – 1st, 2nd and 3rd run through
Another technique that you may try is the first, second, third run through technique.

With this technique, you run through the questions first time round, only answering the questions that you know the answers as soon as you read the question.

The second run through you answer all the questions that you can answer after a little thinking.

Then the third run through should focus on all the remaining questions.

Time Management
It is also important to think about time management throughout the exam. You have 60 minutes in order to answer 60 questions.

My advice would be that if you find that you are spending longer than a minute on a question, or you find that you have had to read the question three times, I would move onto the next question. You do not want to end up spending 2 or even 3 minutes on a question that you will probably end up guessing anyway as you are probably over thinking it, you are better spending the time on questions you can answer confidently.

If you do decide to move on to the next question, make sure you are putting your next answer next to the correct number on the answer sheet.

Question techniques

Here are some exam tips when it comes to answering the questions.

The number one tip for any exam is to. READ THE QUESTION CAREFULLY.

If you can, try to narrow the answers down. Usually you can narrow the answers down as there will often be two answers that you can discard as being incorrect. If you can then just cross the answers off of the question booklet. This will at least narrow it down to a 50/50 at worst.

As mentioned earlier, do not spend too long on a question, if you find that you are spending long then a minute on a question, or you find that you have had to read the question three times or more, I would move onto the next question.

Trust your instincts. In these types of exams you may need to trust your instincts. Think twice about changing your answers. I am not saying that you should never change your answers, just be aware that statistically in these types of exams, people change more correct answers for incorrect answers than the other way round.

And finally, remember each question is individual. Do not panic if you get a run of the same answer, for example, three C's in a row. That does happen in these exams. You should be sure to answer each question individually and not focus on patterns

Prince2 2017 – Practitioner Exam - Guidelines

Exam Structure
- Scenario Based exam – Objective Testing Exam
- 2.5 Hours duration
- 68 Questions
- Pass Mark 55%
- Open Book

Looking at these points in a little more detail

Scenario Based exam – Objective Testing Exam
The Prince2 2017 practitioner exam is a scenario based exam. You will therefore receive a scenario booklet that will contain a 1 to 1 ó page scenario, of which the majority of questions in the exam will be based.

The scenario booklet will also contain some additional material that may be used for a specific question (or part of a question), when this is needed, it will be clearly stated in Bold at the start of the question.

2.5 Hours in duration
The exam is a 2.5 hour exam, this includes any reading time, there is no additional time provided for reading the scenario or additional information. We'll talk about time management in the Exam Strategy section.

68 Questions
The exam will consist of 68 questions in the style of standard multiple choice and matching questions

Pass mark 55%
The pass mark for the exam is 55%, therefore, you will need to score 38 out of 68 to be awarded a pass in this paper.

Open Book
Finally, the exam is an open book exam, this means that you are allowed your Prince2 2017 manual only, you will NOT be allowed any other additional support material outside of the manual. You can write, draw or highlight anything you like in the manual, you are also allowed to tab the pages, to index them. You are NOT, however, allowed to add pages to the manual. Any information you want in the exam will need to be written or drawn directly into the manual.

Topics Covered

The PRINCE2 2017 Practitioner exam will cover the following topics, and will have the following number of questions based on each topic

- Standard Multiple choice (35)
- Matching (33)

Standard Multiple choice

This style of question is very similar to the foundation style questions where you will just pick one answer. This will normally be from 4 options.

The questions relating to the process will have 2 answers that answer correctly the first section of the question, leaving the second part the justification for the first

Once selected – read the answer again with the aim to confirm the whole answer aligns

In this example –

The question is which action should the project manager take and why

The answer is C

The Quality Management Approach contains information regarding compliance and external standards the project must meet, so the project manager should record the information within the Quality Management Approach

Part B answers why Part A is correct

22) The record company must comply with music industry regulations when producing the 'artwork'.

Which action should the project manager take, and why?

A. Record the need to meet this requirement during stage 2, because the 'artwork' will be delivered to the specified quality criteria during stage 3.
B. Record the need to meet this requirement during stage 2, because the product description for the 'artwork' will specify the required quality criteria.
C. Record the requirement in the quality management approach, because compliance with external standards should be addressed when determining the approach to quality.
D. Record the requirement in the quality management approach, because independent quality assurance needs to be planned at the beginning of the project.

PM Guide

Matching

Another style of question that you will encounter is what is known as a Matching question.

They typically involve matching a statement in column 1 with a selection from column 2.

It is important to read the text above the question box, as you can see below from the question relating to quality, the information tells the delegate exactly where to look within the manual by indicating that the headings to be selected are from the Project Product Description

QUALITY

Here are three items of information that will be included in the project product description for the 'album ready for launch'.

Under which heading (A-F) should they be recorded?
Choose only one heading for each item of information. Each heading can be used once, more than once, or not at all.

19) 'Recorded album', 'registered artwork' and 'launch event plan'.	A. Purpose. B. Composition. C. Development skills required. D. Project-level quality tolerances. E. Acceptance method. F. Acceptance responsibilities.
20) The singer will give final approval of the 'artwork'.	
21) The 'artwork' must comply completely with relevant equality legislation.	

Tabbing the Manual

As part of understanding the manual and finding information quickly, it is important to tab your manual to save "page flicking"

Recommended tabs are seen below – the crucial tabs are the most common, the optional being additional and down to individual preference

Crucial	Principles	Theme Business Case	Theme Quality	Theme Quality	Theme Plans	Theme Risk	Theme Change	Theme Progress	Appendix A	Appendix C
	P19	P44	P57	P77	P93	P119	P137	P147	289	P339
	Process Model	SU	DP	IP	CS	MP	SB	CP		
	P158	P165	P179	P195	P215	P235	P245	P259		
	Project Board	Executive	Senior User	Senior Supplier	Project Manager	Team Manager	Project Assurance	Change Authority	Project Support	
	P338	P340	P341	P341	P342	P344	P345	P347	P347	
Optional	Benefits Management Approach	Business Case	Change Control Approach	Checkpoint Report	Communication Management Approach	Configuration Item record	Daily Log	End Project Report	End Stage Report	Exception Report
	P292	P294	P296	P298	P299	P301	P301	P301	P303	P305
	Highlight Report	Issue Register	Issue Report	Lessons Log	Lessons Report	Plan	Product Description	Product Status Account	Project Brief	PID	Project Product Description
	P306	P308	P309	P311	P312	P313	P315	P317	P317	P319	P322
	Quality Management Approach	Quality Register	Risk Management Approach	Risk Register	Work Package					
	P324	P325	P327	P329	P330					

41

PM Guide

Prince2 2017 - Things to add to the manual

As part the Practitioner exam, one of the keys to passing the exam is understanding the manual and where the information is in the most efficient manner, the following are best added to the manual either in the pages at the beginning or at the beginning of each chapter – there is a white space near the chapter no as seen below in the figure from Chapter 6 – Business Case

6 Business case

6.1 The business case theme

Figure 7 - Business Case Theme

Themes

Business Case Theme

Business case –
- Business Case Theme - Chapter 6 – P46-56
- Responsibilities for the Theme – Table 6.1 - P52/53 (Table 6.1)
- Outputs, outcomes and benefits – Figure 6.1 - P46
- Development path – Figure 6.2 - P48/49
- Minimum Requirements – 6.2 – P48

Additional guidance
- Business Case format – 6.3.1 – P43
- Customer\Supplier Business Case – 6.3.3 – P54
- Within a programme – 6.3.4 - P54
- Using an Agile approach – P54

Content of (Appendix A):
- Business case – A.2 P294/295
- Benefits management approach – A.1 - P292/293

Organisation Theme

Organisation
- Organisation - Chapter 7 – Page 58-76
- Minimum Requirements – Chapter 7.2 – P62
- Roles and responsibilities – P62-67 or P338-348 (Appendix C)
- Combining roles – 7.2.1.10 - P67
- Organisation Responsibilities– P68 (Table 7.2.3)

Additional guidance
- Within a programme – P69/70
- In a commercial environment – P71
- Using an Agile approach – P72

Content of
- Appendix C – Roles & Responsibilities C1-C9
- Communication management approach – A5 - P299

Quality Theme

Quality
- Quality Theme - Chapter 8 – Page 78
- Roles & Responsibilities – P82/83 (Table 8.1)
- Quality planning and control – P79/80
- Quality Assurance – 8.3.5 – P84
- In-Process & Appraisal Reviews – 8.3.13 – P87-88
- Quality Review Technique – P89-92
- Minimum Requirements – 8.2 – P80-82

Content of (Appendix A):
- Quality Management Approach – A.22 - P324
- Quality Register – A.23 - P325
- Project Product Description – A.21 - P322
- Product Description – A.17 - P315/316

Plans Theme

Plans
- Plans Theme Chapter 9 – Page 94-118
- Minimum Requirements – 9.2 – P97\98
- Project \Stage\Team Plan – 9.2.1.1, 2. 3 – P99/100
- Roles & Responsibilities – P101 (Table 9.1)
- Planning approach – P102-114 (Diagram – P102 – Figure 9.2, P106 – Figure 9.6)
- Stage considerations – P103-105
- Product-Based Planning – P350 (Appendix D)

Additional guidance
- Within a programme – 9.3.2 - P115
- From a supplier perspective – 9.3.4 - P115
- Using an Agile approach – 9.3.3 - P115

Content of (Appendix A)
- Plan (Project, stage or Team) – A.16 - P313/314

PM Guide

Key Definitions
- Internal product – Something created by the project (within the scope and control of the project)
- External product – Something NOT created by the project, but is needed. It either already exists (Existing, current, previous, available), or is being created by another project.

Risk Theme
Risk
- Risk Theme – Chapter 10 – Page 1200136
- Minimum Requirements – 10.2 – P121-122
- Roles & Responsibilities– 10.2.1 - P122 (Table 10.1)
- Alignment with Org & other polices\process – 10.3.1 – P123
- Risk procedure (Identify, Assess, Plan, Implement, Communicate) – 10.3.2 - P126-135
- Risk responses – P132 (Table 10.3)
- Risk Owner/Actionee – P134 (10.4.4)

Content of (Appendix A)
- – Risk management approach – A.24 - P327/328
- – Risk register – A.25 - P329

Change Theme
Change
- Change Theme – Chapter 11 – Page 138
- Minimum Requirements – 11.2 – P140
- Roles & Responsibilities – P141 (Table 11.2)
- Types of issues (RFC, Off-spec, Problem/concern) – P138 (Table 11.1)
- Issue/Change control procedure – P144 (Figure 11.1)

Content of (Appendix A):
- Change control approach – A.3 - P296/297
- Issue register – A.12 - P308
- Issue report – A.13 - P309/310

Progress Theme
Change –
- Progress – Chapter 12 – Page 148

PM Guide

- Minimum Requirements – 12.2 - P148
- Roles & Responsibilities – Table 12.2 - P154
- Tolerances – Table 12.1 - P149
- Exceptions – 12.2.3 – P 153

Basics
- Tolerance – Permissible deviation above or below an objective before the need to escalate
- Exception – Forecasted deviation outside of tolerance
- Delegation of authority – One management level below
- Reporting and escalations – to the next management level up
- Escalation Path – never skip\by[ass a management level (See figure 12.1 on P150)

Content of (Appendix A):
- Highlight report – A.11 - P306/307
- Checkpoint report – A4 - P298
- End stage report – A.9 - P303/304
- End project report – A.8 - P301/302
- Exception Report – A.10 – P305
- Work package – A.26 - P330/331

Processes
Starting Up a Project Process (SU)
Starting Up a Project Process
- Starting up a project – Chapter 14 – Page 166
- Purpose & Objective – 14.1 & 14.2 – P166\67
- Tailoring guidelines – P177/178 (Section 14.5)

Additional Guidance
- Simple Project – 14.5.4.1 – P178
- Agile Delivery Approach – 14.5.4.2 – P178
- Supplier Perspective – 14.5.4.3 – P78
- Within a Programme – 14.5.4.4 – P78

Content of (Appendix A):
- Project Brief – A.19 - P317/318

45

PM Guide

Directing a Project Process (DP)
Directing a Project Process
- Directing a Project Process – Chapter 15 – Page 180
- Purpose & Objective – 15.1 & 15.2 – P180
- Context – 15.3 – P180/P181
- Tailoring Roles – 15.5.2 – P192

Additional Guidance
- Simple Project – 15.5.3.1 – P192
- Agile Delivery Approach – 15.5.3.2 – P192
- Supplier Perspective – 15.5.3.3 – P193
- Within a Programme – 15.5.3.4 – P193

Initiating a Project Process (IP)
Initiating a Project Process
- Initiating a Project – Chapter 16 – Page 196
- Purpose & Objective – 16.1 & 16.2 – P196
- Tailoring Guidelines – 16.5 - P212/213

Additional Guidance
- Simple Project – 15.5.3.1 – P192
- Agile Delivery Approach – 15.5.3.2 – P192
- Supplier Perspective – 15.5.3.3 – P193
- Within a Programme – 15.5.3.4 – P193

Controlling a Stage Process (CS)
Controlling a Stage Process
- Controlling a Stage – Chapter 17 – Page 216
- Purpose & Objective – 17.1 & 17.2 – P216
- Tailoring Guidelines – 17.5 – P231

Additional Guidance
- Simple Project – 17.5.4.1 – P232
- Agile Delivery Approach – 17.5.4.2 – P232
- Supplier Perspective – 17.5.4.3 – P233
- Within a Programme – 17.5.4.4 – P233

PM Guide

Managing a Product Delivery (MP)
Managing Product Delivery
- Managing Product Delivery – Chapter 18 – P236
- Purpose & Objective – 18.1 & 18.2 – P236
- Tailoring Guidelines – 18.5.2 – P242

Contents of (Appendix A)
- Work Package – A.26 – P330
- Checkpoint Report – A.4 – P298

Managing a Stage Boundary Process (SB)
Managing a Stage Boundary Process
- Managing A Stage Boundary – Chapter 19 – Page 246
- Purpose & Objective – 19.1 & 19.2 – P246
- Tailoring Guidelines – 19.5.2 – 256\7

Contents of (Appendix A)
- End Stage Report – A9 – P303
- Plan – A.16 – P313
- Lessons Report – A.15 – P312
- Exception Plan – A.16 – P313

Closing a Project Process (CP)
Closing a Project Process
- Closing a Project Process – Chapter 20 – P260
- Purpose & Objective – 20.1 & 20.2 – P260
- Planned closure – 20.4.1 – P262
- Premature Closure – 20.4.2 – P263

Contents of (Appendix A)
- End Project Report – A.8 – P301
- Lessons Report – A.15 – P312

www.ingramcontent.com/pod-product-compliance
Lightning Source LLC
Chambersburg PA
CBHW071151220526
45468CB00003B/1022